Contents

How to use this book

Each page has a title telling you what it is about.

Instructions look like this. Always read these carefully before starting.

This shows you how to set out your work. The first question is done for you.

Read these word problems carefully. Decide how you will work out the answers.

Sometimes there is a **Hint** to help you.

Sometimes you need materials to help you.

This means you must decide how to lay out your work and show your workings.

This shows that the activity is an **Explore**. Work with a friend.

Ask your teacher if you need to do these.

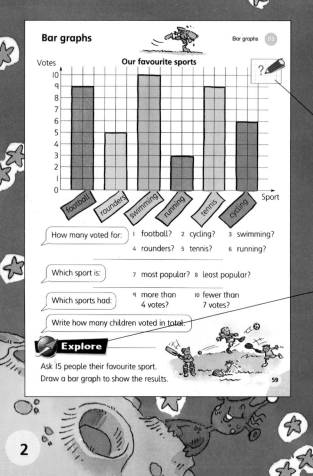

Centimetres (cm)

Find one of each object.   Estimate its length in centimetres.   Use a ruler to measure it.

1

1. estimate 1 0 cm
   length 7 cm

2

3

4

5

6

7

Choose the nearest length.

8

1 cm
4 cm
10 cm

8. 4 cm

9

8 cm
2 cm
18 cm

10

1 cm
50 cm
7 cm

11

10 cm
1 cm
100 cm

Metres (m)

Choose the nearest length.

1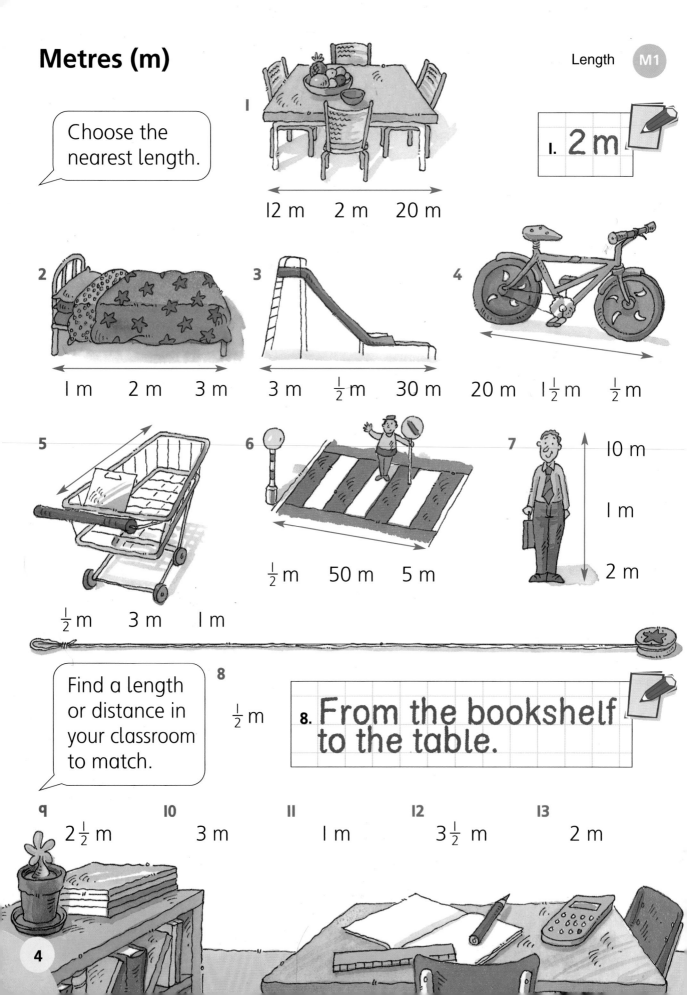

12 m   2 m   20 m

1. **2 m**

2

1 m   2 m   3 m

3

3 m $\frac{1}{2}$ m   30 m

4

20 m $1\frac{1}{2}$ m $\frac{1}{2}$ m

5

$\frac{1}{2}$ m   3 m   1 m

6

$\frac{1}{2}$ m   50 m   5 m

7

10 m

1 m

2 m

Find a length or distance in your classroom to match.

8 $\frac{1}{2}$ m

8. **From the bookshelf to the table.**

9 $2\frac{1}{2}$ m   **10** 3 m   **11** 1 m   **12** $3\frac{1}{2}$ m   **13** 2 m

4

Metres (m)

Write the number of metres.

1   1 km

1. 1 km = 1 0 0 0 m

2   1 km 100 m

3   1 km 500 m

4   2 km

5   2 km 500 m

6   1 km 700 m

7   1 km 900 m

8   3 km

9   2 km 100 m

10   10 km

Explore

You need to run 1 km or more for charity.

Which routes from school could you run?

beach 750 m

museum 400 m

video shop 500 m

200 m football ground

school

350 m forest

100 m playground

Centimetres (cm) and metres (m)

Write how many centimetres.

1 3 m

1. 3m = 300cm

2 1 m

3 2 m 30 cm

4 1 m 40 cm

5 3 m 25 cm

6 1 m 50 cm

7 2 m 15 cm

8 2 m 33 cm

9 1 m 5 cm

Ⓔ Choose 2 snakes. Write their total length. Repeat 5 times.

Write how many metres and centimetres.

10 248 cm

10. 2m 48 cm

11 292 cm

12 475 cm

13 332 cm

14 127 cm

15 255 cm

16 304 cm

17 150 cm

18 111 cm

Ⓔ Write each length in metres using a decimal point (for example 2·48 m).

Centimetres (cm) and metres (m)

Each giraffe grows 25 cm more. Write the new height.

1. $3\,m\,15\,cm + 25\,cm = 3\,m\,40\,cm$

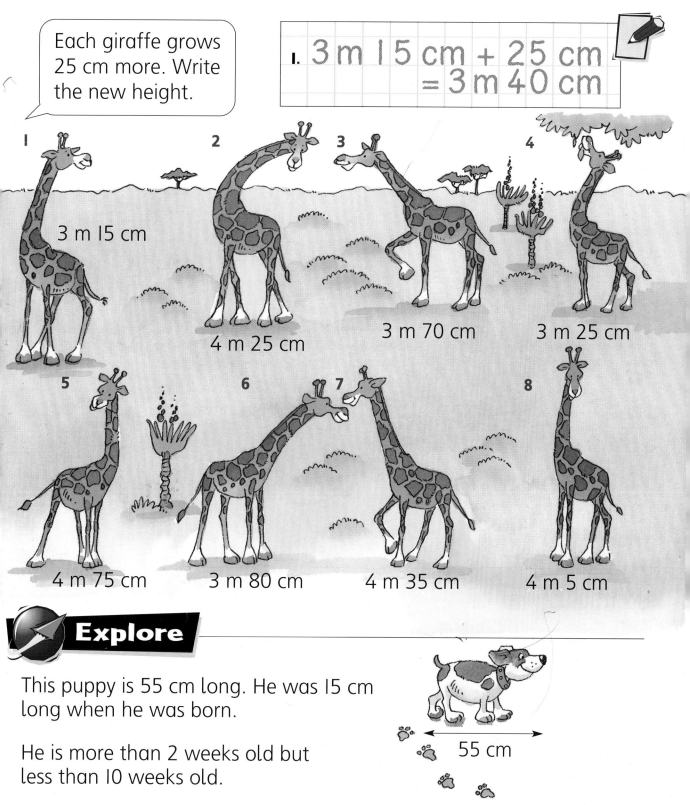

1. 3 m 15 cm

2. 4 m 25 cm

3. 3 m 70 cm

4. 3 m 25 cm

5. 4 m 75 cm

6. 3 m 80 cm

7. 4 m 35 cm

8. 4 m 5 cm

Explore

This puppy is 55 cm long. He was 15 cm long when he was born.

55 cm

He is more than 2 weeks old but less than 10 weeks old.

He has grown the same amount each week.

How old could he be?

How much did he grow each week?

7

Centimetres (cm) and metres (m)

How much taller is each sister?

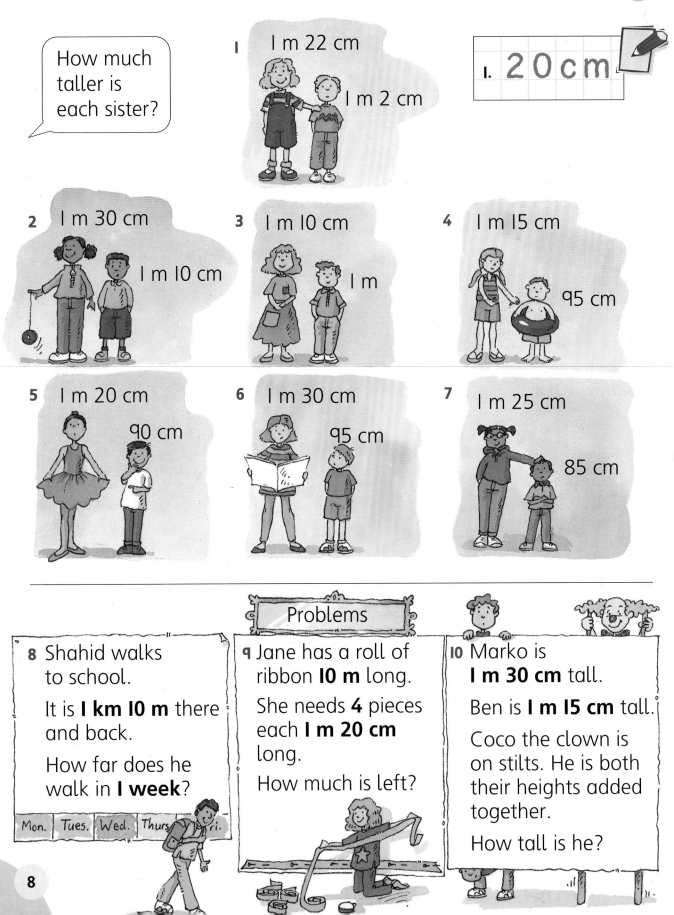

1 1 m 22 cm / 1 m 2 cm

I. 20 cm

2 1 m 30 cm / 1 m 10 cm

3 1 m 10 cm / 1 m

4 1 m 15 cm / 95 cm

5 1 m 20 cm / 90 cm

6 1 m 30 cm / 95 cm

7 1 m 25 cm / 85 cm

Problems

8 Shahid walks to school.

It is **1 km 10 m** there and back.

How far does he walk in **1 week**?

Mon. Tues. Wed. Thurs. Fri.

9 Jane has a roll of ribbon **10 m** long.

She needs **4** pieces each **1 m 20 cm** long.

How much is left?

10 Marko is **1 m 30 cm** tall.

Ben is **1 m 15 cm** tall.

Coco the clown is on stilts. He is both their heights added together.

How tall is he?

Quarter past, half past, quarter to

Which clocks match?

I. a and j

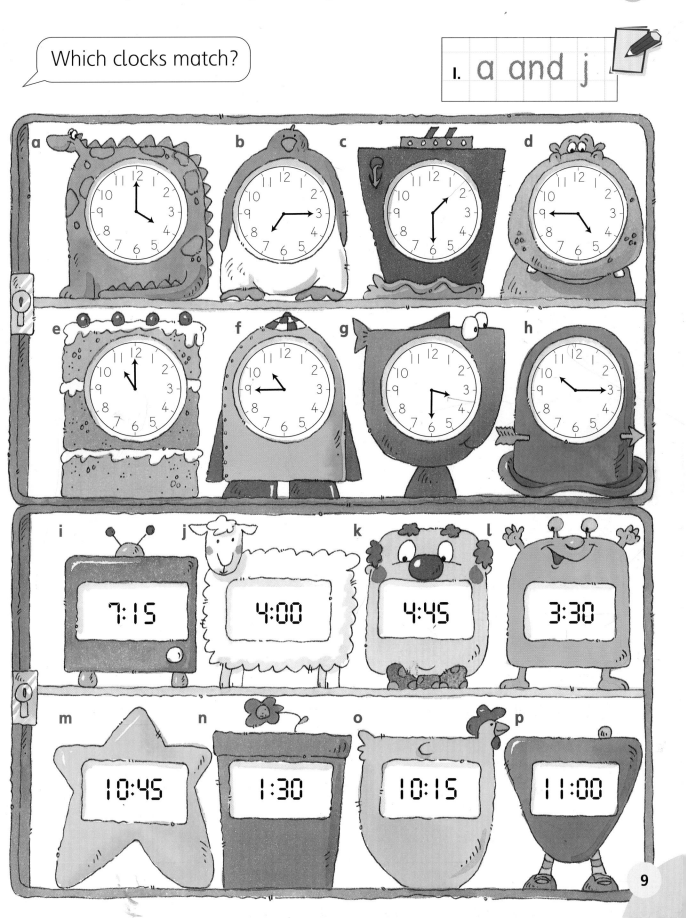

5 minutes

Write how many minutes are coloured.

1. 30 minutes

2

3

4

5

6

7

8

q

Write how many minutes between each pair of clocks.

10

11

12

13

14

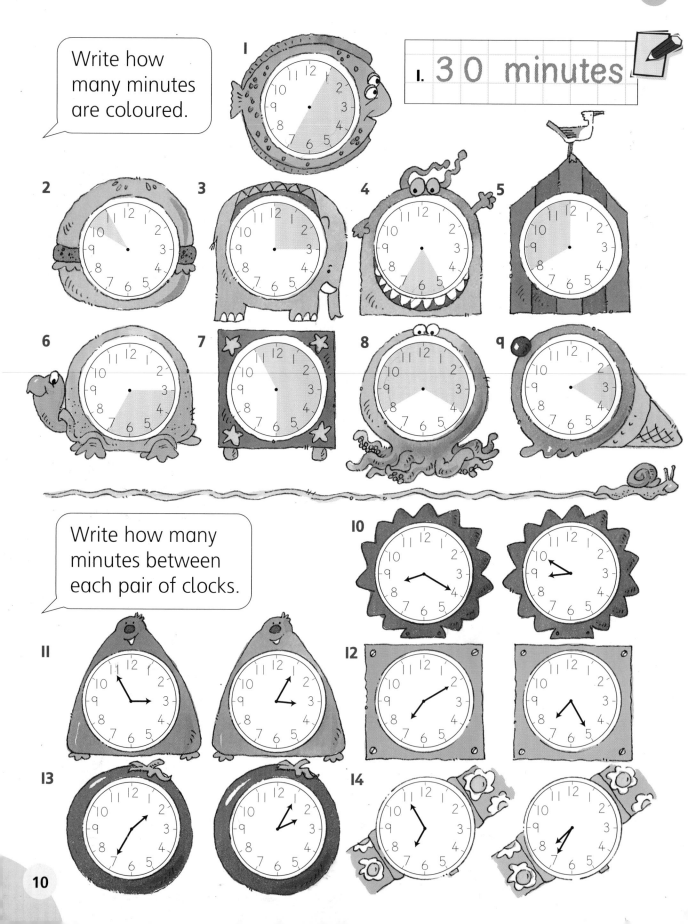

5 minutes

Write the times every 5 minutes to the next hour.

I. 8:45
    8:50
    8:

1 8:40

2 9:30

3 6:00

Problems

4

BUS STOP

It is **quarter past 10**.

Dan's bus comes in half an hour.

What time will it come?

5

BUZZ WORD

Buzz Word starts on TV at **3:45**.

It ends at **4:15**.

How long is the programme?

6 Ann is meeting Shonka and Jan outside the school disco at **6:30**.
It takes **45 minutes** to get there.
When should she leave her house?

DISCO 6:30

Write the time shown on each clock.

I. **ten past 6**

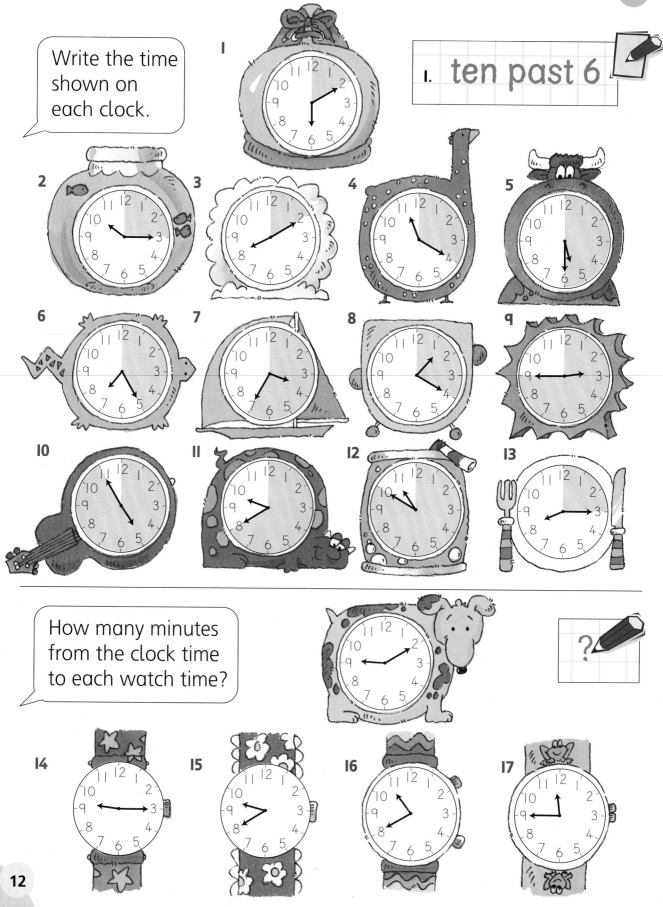

1

2

3

4

5

6

7

8

9

10

11

12

13

How many minutes from the clock time to each watch time?

?

14

15

16

17

5 minutes

Write the time shown on each clock.

1 **5:05**

I. five past 5

2 **2:35**

3 **8:15**

4 **3:25**

5 **6:50**

6 **5:20**

7 **10:10**

8 **7:45**

9 **1:40**

10 **4:55**

11 **9:30**

12 **11:10**

13 **12:05**

Write digital times to match.

14 quarter past 5

14. **5:15**

15 twenty past 8

16 half past 10

17 five past 7

18 twenty to 12

19 ten past 6

20 ten to 9

21 five to 3

22 twenty to 2

23 twenty-five to 4

Write the time 10 minutes later.

1 2:40

1. ten to 3

2 6:15

3 10:25

4 4:45

5 8:05

6 9:10

7 5:45

8 11:50

9 12:55

🄔 Write each time 20 minutes earlier.

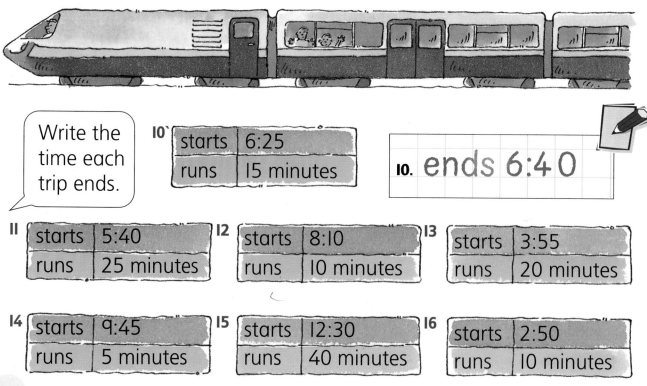

Write the time each trip ends.

10

starts	6:25
runs	15 minutes

10. ends 6:40

11

starts	5:40
runs	25 minutes

12

starts	8:10
runs	10 minutes

13

starts	3:55
runs	20 minutes

14

starts	9:45
runs	5 minutes

15

starts	12:30
runs	40 minutes

16

starts	2:50
runs	10 minutes

Millilitres (ml)

Write **more than**, **less than** or **equal to** 1 litre.

1
30 ml

I. less than 1l

2
250 ml

3
2000 ml

4
330 ml

5
750 ml

6
1000 ml

7
3000 ml

8
1500 ml

q
660 ml

Choose the nearest capacity.

10
50 ml
5 ml
15 ml

10. 15 ml

11
2500 ml
250 ml
25 ml

12
1000 ml
50 ml
5 ml

13
50 ml
5 ml
300 ml

14
300 ml
1500 ml
15 ml

15

Millilitres (ml)

Write how many millilitres of water in each container.

1 | 1 litre

I. 300 ml

2 | 1 litre

3 | 1 litre

4 | 1 litre

5 | 1 litre

6 | 1 litre

7 | 1 litre

8 | 1 litre

q | 1 litre

Write how many millilitres each set of containers holds.

10 1 l

10. 1 l = 1000 ml

11 $\frac{1}{2}$ l   3 l

12 2 l

13 2 l $\frac{1}{2}$ l

14 $\frac{1}{2}$ l

15 3 l

16 6 l

17 4 l

18 $\frac{1}{2}$ l $\frac{1}{2}$ l   2 l

Millilitres (ml)

Write the amounts in order, smallest to largest.

50 ml
200 ml

Jam
900 ml

milk
2 l

3000 ml

650 ml

2500 ml

1 l

½ l

50 ml

200 ml

4 l

1500 ml

Problems

1 Jen drank **350 ml** from her **1 litre** bottle.

How much is left?

2 A spoon holds **5 ml**. How many spoonfuls of water to make ½ **l**?

3 Minnie's cup holds **150 ml** of juice.

How many cupfuls are there in a **1 litre** carton?

Explore

Use 2 sheets of A4 paper.

Make 2 cylinders as shown.

Which holds the most?

Work with a friend.

Lentils or rice

Hours and minutes

Write how many minutes long.

1.
I hour

I. 1 hour = 60 minutes

2. I hour 30 minutes

3. I hour 20 minutes

4. I hour 50 minutes

5. 2 hours

6. I hour 40 minutes

7. I hour 45 minutes

8. 2 hours 10 minutes

9. I hour 55 minutes

10. I hour 10 minutes

Write how many hours.

II. 1 hour 15 minutes

11. 75 minutes

12. 85 minutes

13. 90 minutes

14. 100 minutes

15. 80 minutes

16. 120 minutes

17. 130 minutes

18. 95 minutes

19. 180 minutes

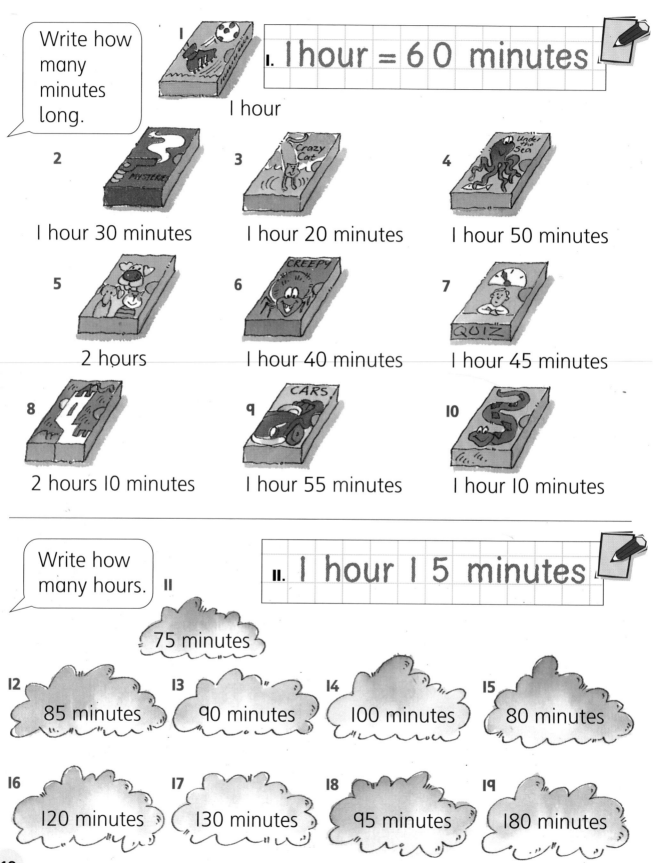

Days and hours

The letters in each day are jumbled.

Write the days in the correct order.

Monday
Tu

Write how many days.

Breaking Records

1. more than 1 day

1. 30 hours

2. 40 hours

3. 70 hours

4. 50 hours

5. 20 hours

6. 80 hours

7. 90 hours

Seconds and minutes

Write each race time in minutes.

1. 65 seconds

1. I minute and 5 seconds

2. 75 seconds

3. 85 seconds

4. 110 seconds

5. 120 seconds

6. 90 seconds

7. 130 seconds

Write the answers.

Calculator

?

8 How many seconds in a day?

9 How many hours in a week?

10 How many minutes in Saturday and Sunday?

Weekend Break

☻ Make up some questions of your own like these and answer them.

Grams (g) and kilograms (kg)

Find the objects and place them on the balance.

How many 100 g weights to tip the balance?

1. 4

A balance
Weights

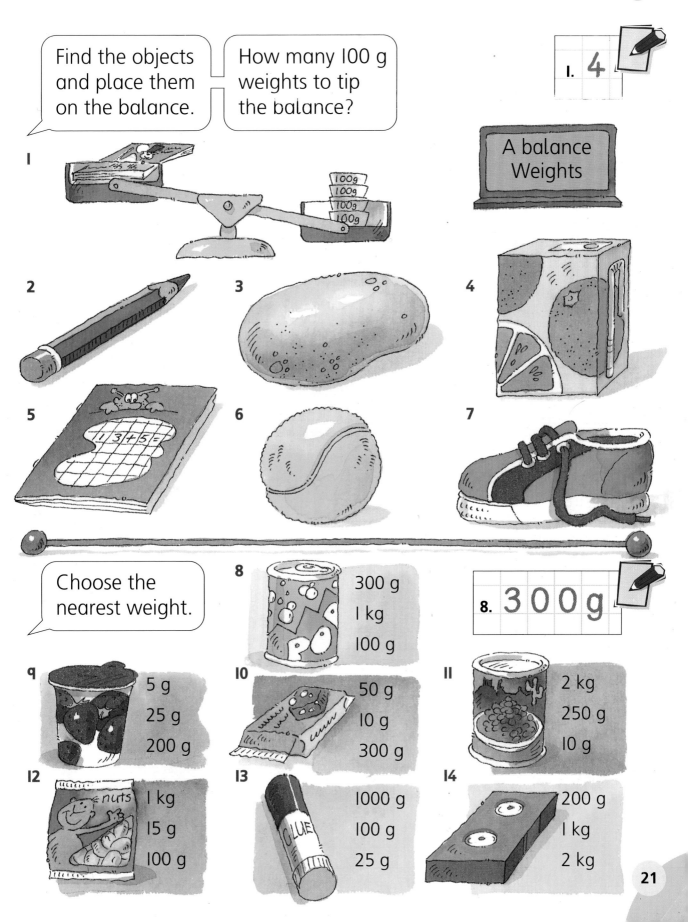

Choose the nearest weight.

8.
300 g
1 kg
100 g

8. 300 g

9.
5 g
25 g
200 g

10.
50 g
10 g
300 g

11.
2 kg
250 g
10 g

12.
1 kg
15 g
100 g

13.
1000 g
100 g
25 g

14.
200 g
1 kg
2 kg

21

Write the weight of each alien's travel bag.

1.

1. 4 kg 500 g

1
4 kg 5

2
1 kg 2

3
7 kg 8

4
0 kg
1 kg

5
0 1 2 3 4 kg

6
3 kg
4 kg

7
0 kg 4
2

How many of each to make 1 kg?

8
500 g

8. 2

9
200 g

10
100 g

11
250 g

12
50 g

Explore

You are going to Mars.

You can take up to 2 kg in your rucksack.

What could you take?

350 g

200 g

125 g

200 g

700 g

250 g

100 g

600 g

50 g

Grams (g) and kilograms (kg)

Write each weight in grams.

I. 1000g

1. 1 kg

2. ½ kg

3. 2 kg

4. 1½ kg

5. 1 kg 300 g

6. ¾ kg

7. 5 kg

Problems

8 Each plum weighs **40 g**.

How much do **6** plums weigh?

9 Abdul's mum bought **1 kg** of tomatoes.

Each tomato weighs **50 g**.

How many tomatoes did she buy?

10 The pineapples weigh ½ **kg** each.

How much do **10** pineapples weigh?

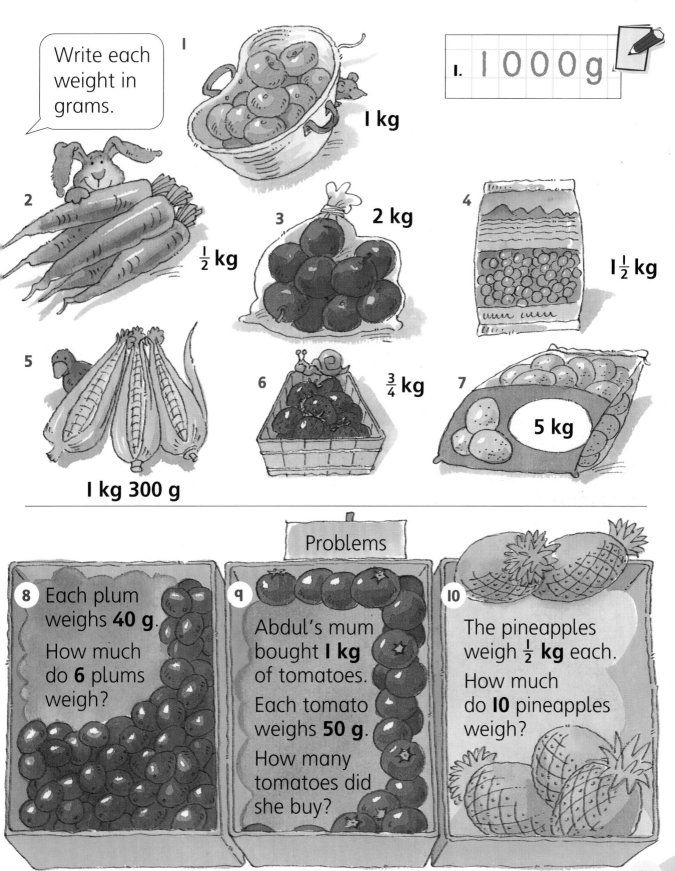

The letters in each month are jumbled.

Write the months in the correct order.

January
Fe

Write how many years and months old.

1 16 months

1. I year and 4 months

2 26 months

3 30 months

4 20 months

5 23 months

Days

Write how many days.

I. 2 weeks = 14 days

1 USA 2 weeks

2 Australia 6 weeks

3 Wales 2 weeks 2 days

4 India 2 weeks 5 days

5 Devon 2 weeks 1 day

6 France 2 weeks 6 days

7 Disney World 3 weeks

8 Visit Santa in Lapland 1 week 5 days

Write how many days to each birthday.

q. 7 days

9 10 July — 3 July

10 2 Sept — 12 Sept

11 15 April — 4 April

12 2 March — 17 March

13 7 April — 21 April

14 25 Dec — 12 Dec

15 6 August — 18 August

Days, months and years

Write the missing number.

I. 7 days in I week

I	days in I week	**2**	days in 4 weeks
3	years in a decade	**4**	months in 2 years
5	weeks in a year	**6**	days in a year
7	weeks in half a year	**8**	weeks in 2 years
9	months in a decade	**10**	years in a century

Explore

How many days old are you?

Guess first.

Then ... use a calculator, pencil and paper to work it out.

Shape names

triangle ⟩ square ⟩ rectangle ⟩ pentagon ⟩ hexagon ⟩ octagon

For each shape, write the number of sides and its name.

1

I. 5 sides
pentagon

2

3

4

5

6

7

8

9

10

Draw 2 different pentagons. Label them.

Draw 2 different hexagons then 2 different octagons.

pentagon

Ruler

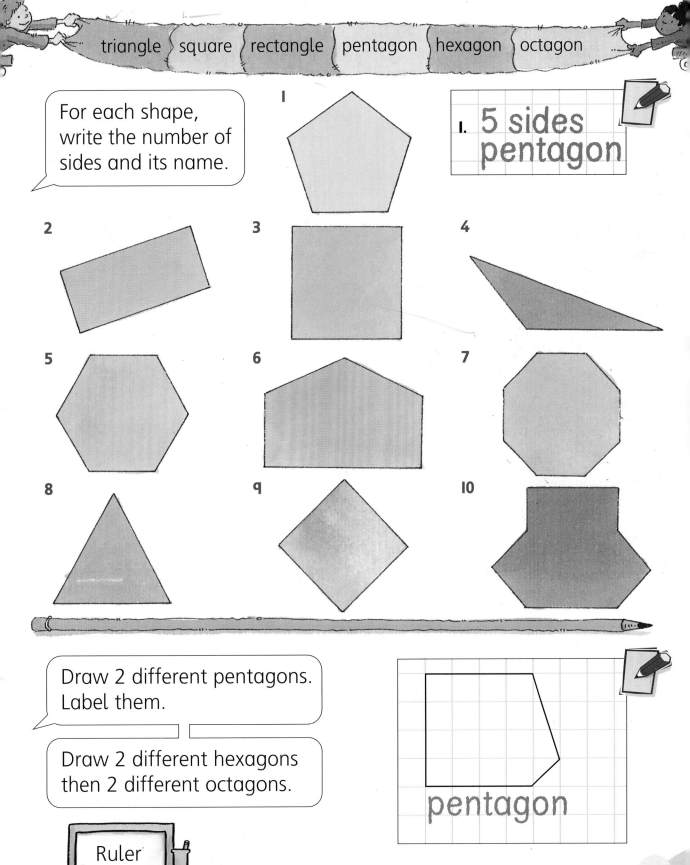

Quadrilaterals

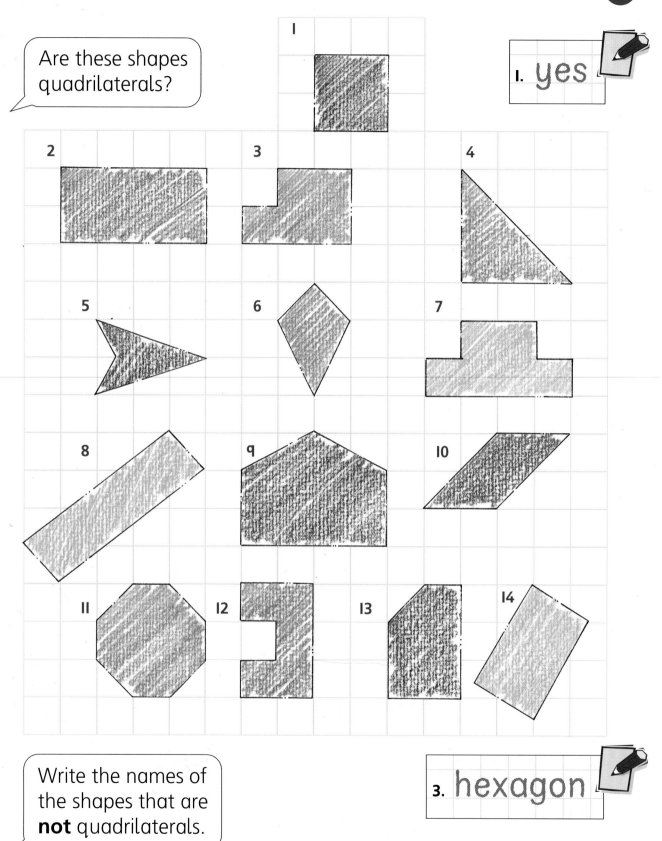

Are these shapes quadrilaterals?

I. yes

Write the names of the shapes that are **not** quadrilaterals.

3. hexagon

Shape names

Draw these shapes and name them.

1.

I.
hexagon

2

3

4

5

6

7

8

9

℮ Draw 4 more shapes and name them.

Draw 6 different quadrilaterals.

Squared paper, ruler

Shape names

Fold a sheet of paper and make these cuts.

Stick the shape in your book and name it.

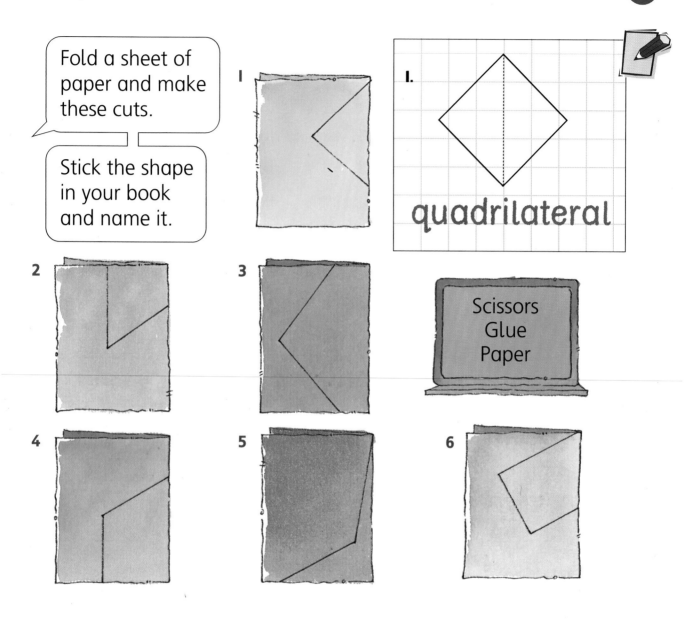

1.

quadrilateral

Scissors
Glue
Paper

Explore

Fold sheets of paper.

Make two straight cuts.

Draw the fold-line on the opened shapes.

Make a display.

Lines of symmetry

Are these lines of symmetry?

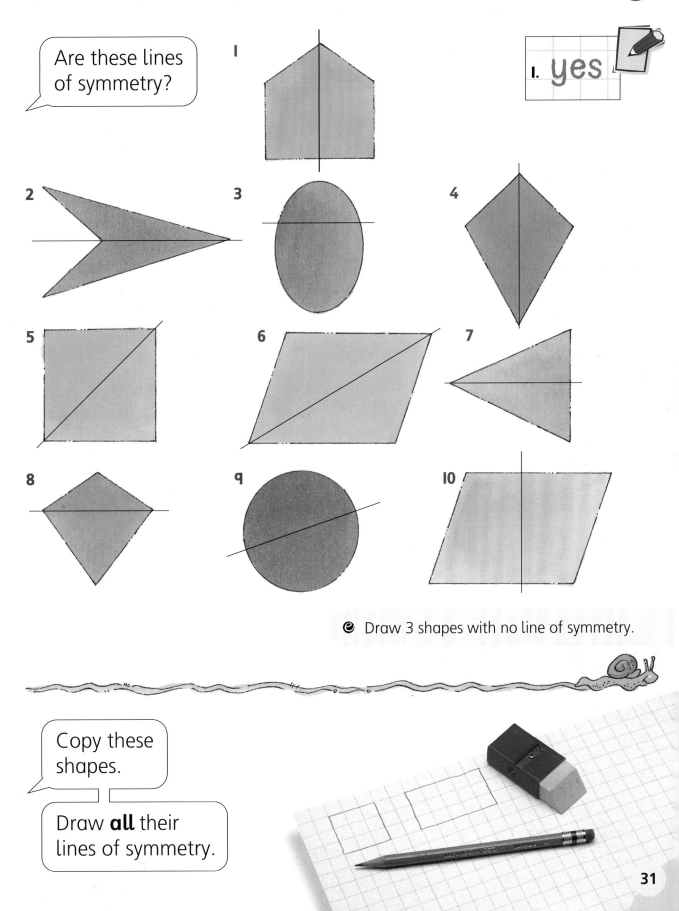

1. yes

● Draw 3 shapes with no line of symmetry.

Copy these shapes.

Draw **all** their lines of symmetry.

Lines of symmetry

Copy these shapes.

Draw **all** their lines of symmetry.

2 cm squared paper
Scissors

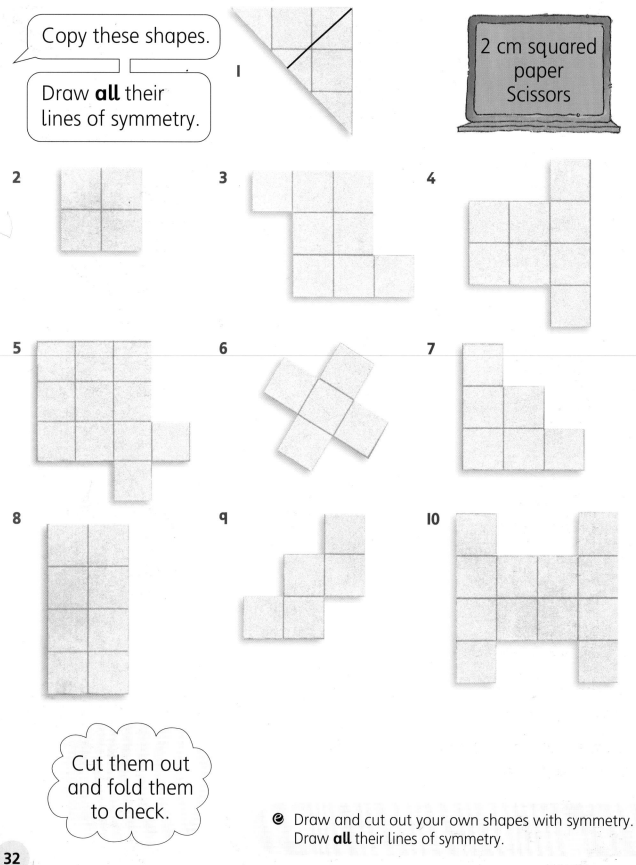

Cut them out and fold them to check.

● Draw and cut out your own shapes with symmetry.
Draw **all** their lines of symmetry.

Lines of symmetry

Write how many lines of symmetry each badge has.

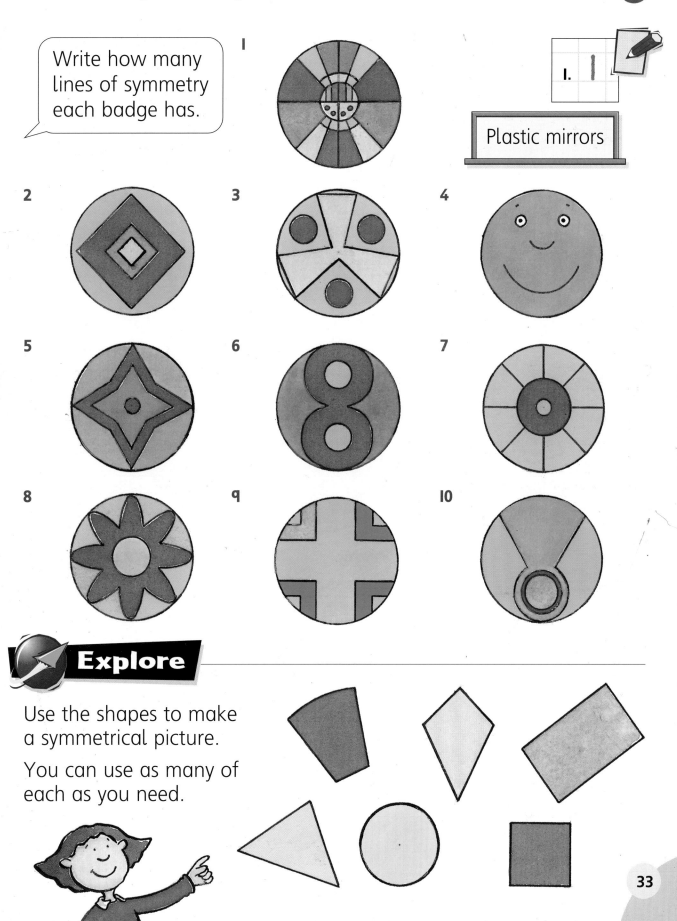

Plastic mirrors

I.

1

2

3

4

5

6

7

8

9

10

Explore

Use the shapes to make a symmetrical picture.

You can use as many of each as you need.

North, south, east, west

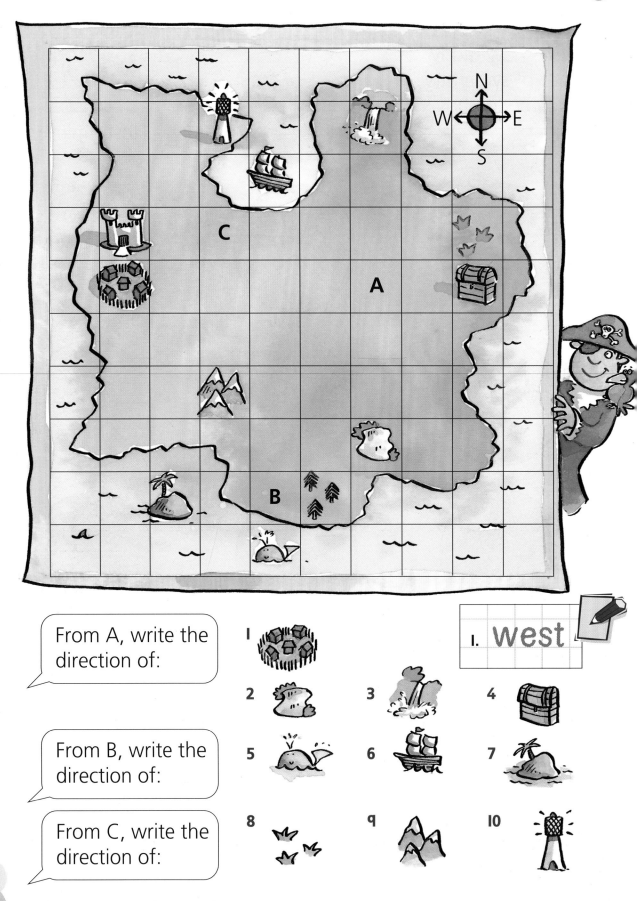

From A, write the direction of:

1

I. **west**

2

3

4

From B, write the direction of:

5

6

7

From C, write the direction of:

8

9

10

34

North, south, east, west

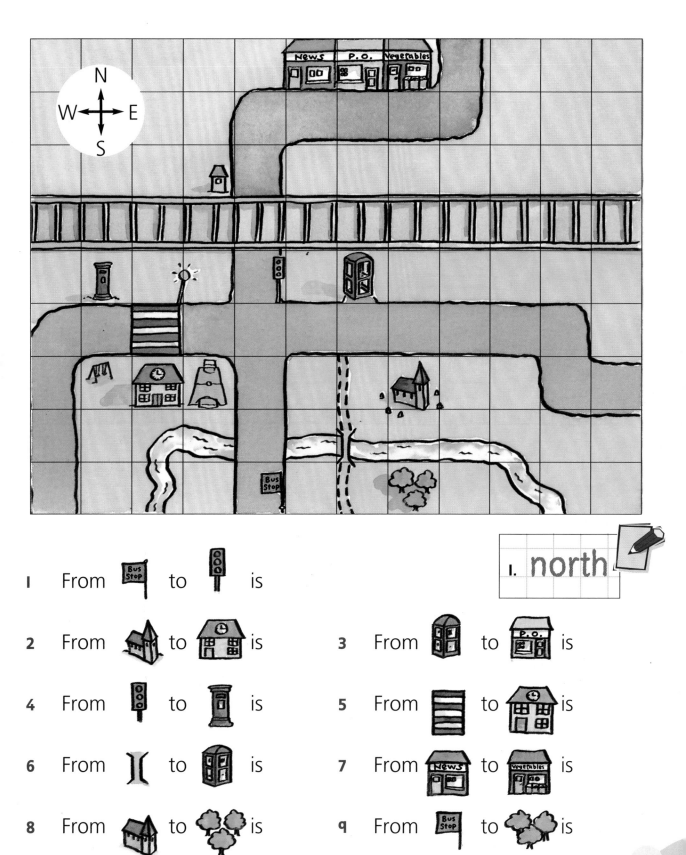

1 From to is

I. north

2 From 🏛 to 🏠 is   3 From 📞 to 🏤 is

4 From 🚦 to 🏛 is   5 From ▤ to 🏠 is

6 From) (to 📞 is   7 From News to Vegetables is

8 From 🏛 to 🌳🌳 is   9 From Bus Stop to 🌳🌳 is

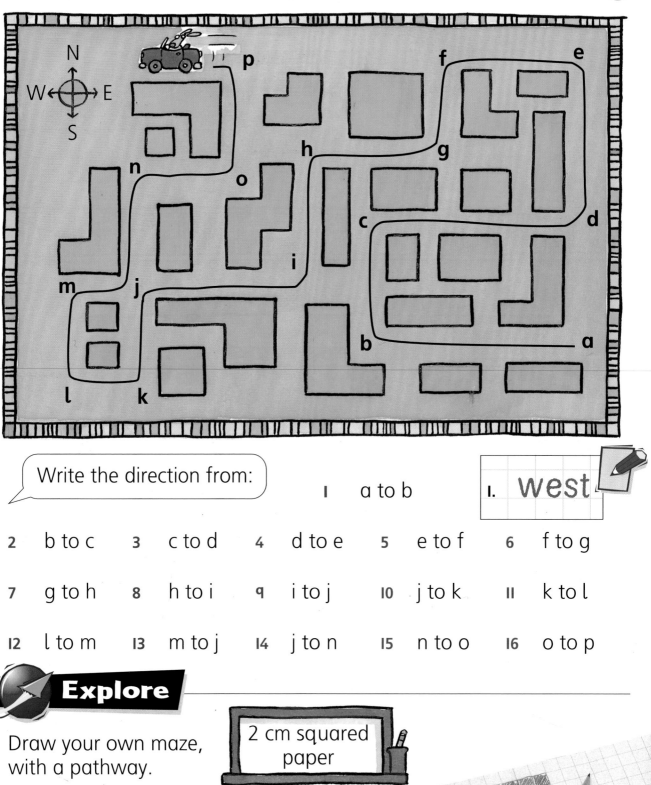

Write the direction from:

I	a to b

I. **west**

2	b to c	3	c to d	4	d to e	5	e to f	6	f to g
7	g to h	8	h to i	9	i to j	10	j to k	11	k to l
12	l to m	13	m to j	14	j to n	15	n to o	16	o to p

Explore

Draw your own maze, with a pathway.

2 cm squared paper

Write the direction for each part of the journey.

Prisms

Are these prisms?

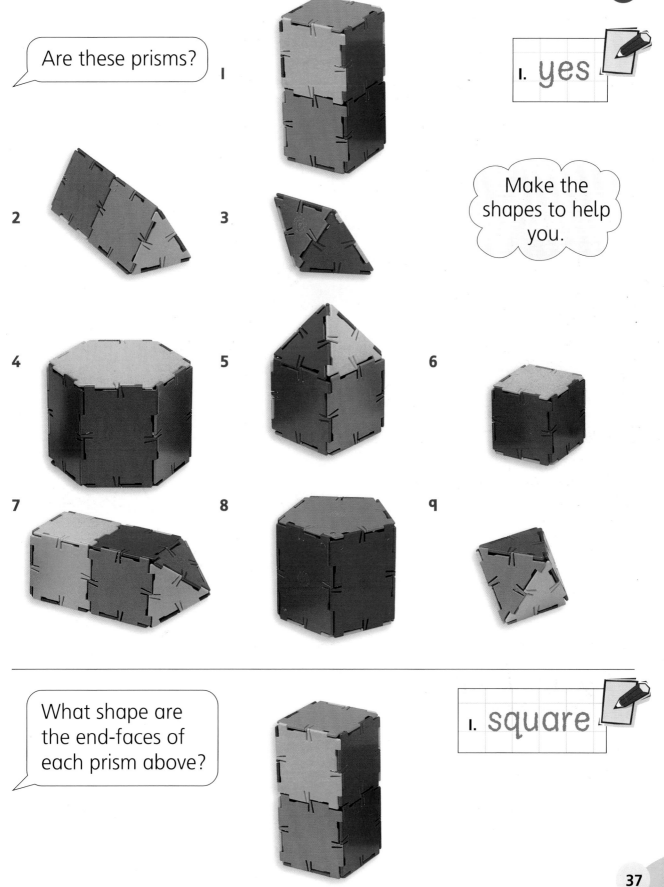

Make the shapes to help you.

1. **yes**

1

2

3

4

5

6

7

8

9

What shape are the end-faces of each prism above?

1. **square**

Prisms

3-d shape **S4**

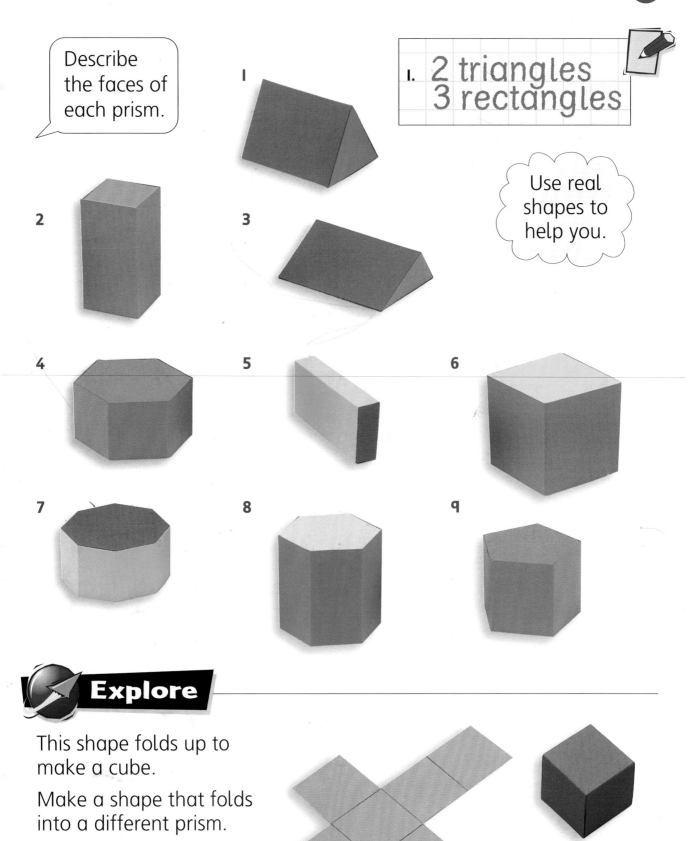

Describe the faces of each prism.

1. 2 triangles
3 rectangles

Use real shapes to help you.

Explore

This shape folds up to make a cube.

Make a shape that folds into a different prism.

38

Names of shapes

Write the name of each shape.

I

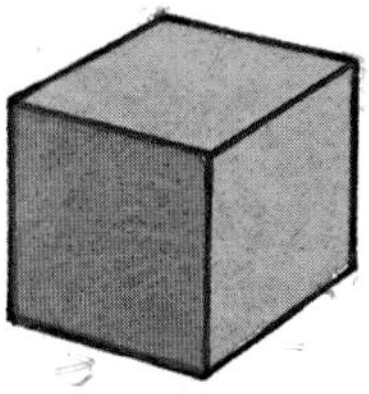

I. cube

2

3

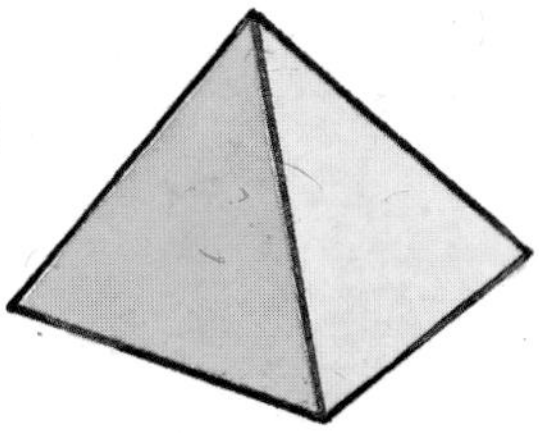

4

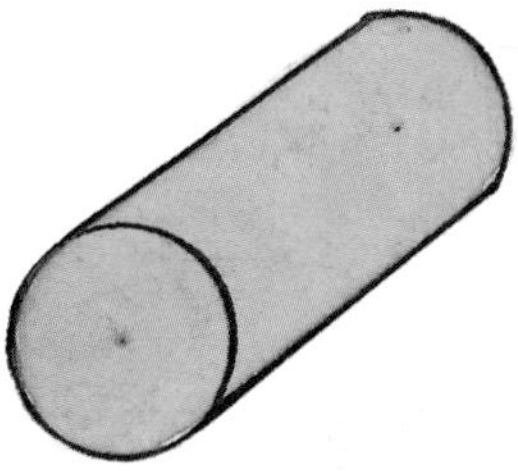

5

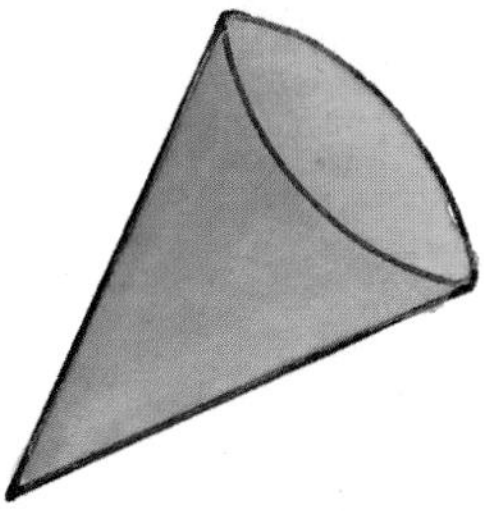

6

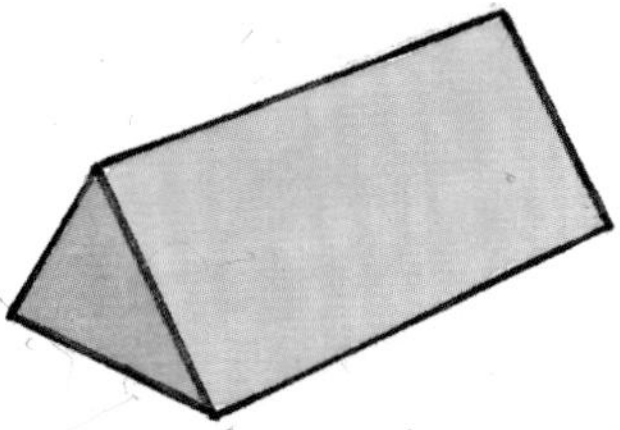

7

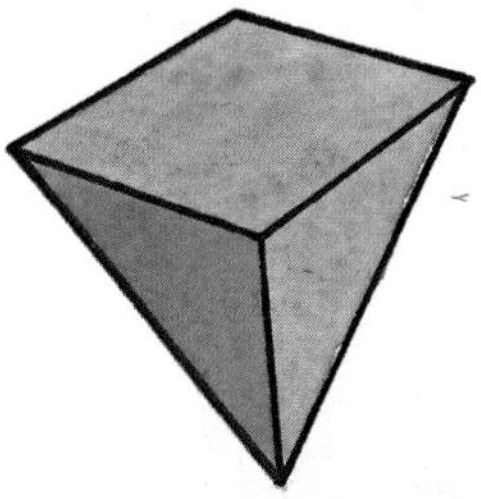

8

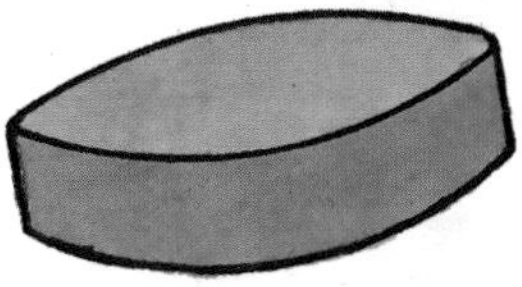

q

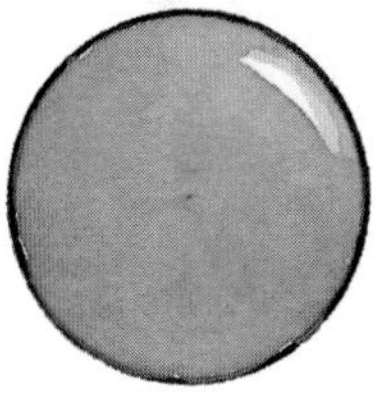

10

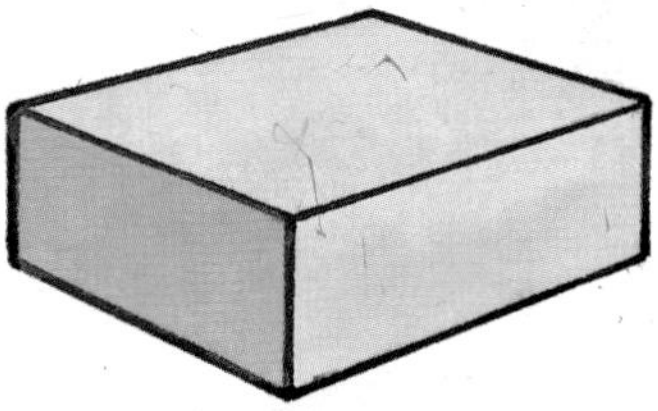

cube

pyramid

cuboid

prism

cone

sphere

cylinder

Write the shapes with:

II a curved face

II. 4, 5, 8, q

12 6 faces

13 2 circular faces

14 6 vertices

15 all faces the same

16 5 vertices

17 I circular face

Sorting shapes

> For each shape, write the number of faces, vertices and edges.

I.   6 faces
      8 vertices
    1 2 edges

> Use real shapes to help you.

1

2

3

4

5

6

7

8

9

10

Explore

Find or make some differently shaped boxes.

Label each box with the number of faces, vertices and edges.

6 faces
8 vertices
12 edges

Sorting shapes

a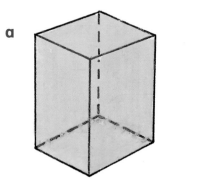

b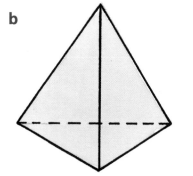

c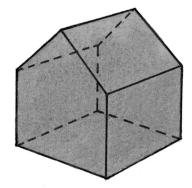

d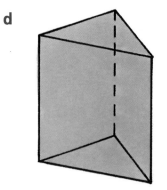

e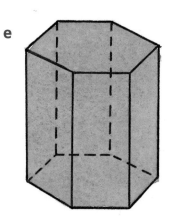

f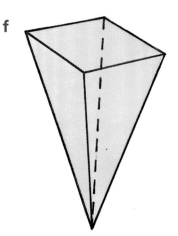

Copy and complete these tables.

Use real shapes to help you.

shape				a		
faces	4	5	5	6	7	8

shape						
edges	6	8	9	12	15	18

shape						
vertices	4	5	6	8	10	12

Sorting shapes

Write how many:
– cubes
– cylinders
– cuboids
– prisms.

 **Explore**

Collect several food packets which are prisms.

Sort them into different types.

Investigate the capacities of different prisms.

42

Right angles

Are these right angles?

1.

I. No

2

3

4

A right angle measure

5

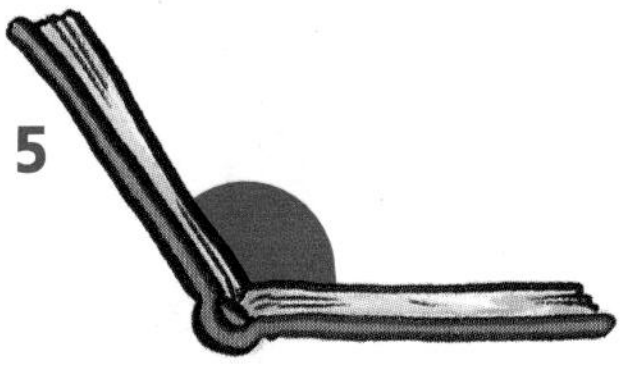

6

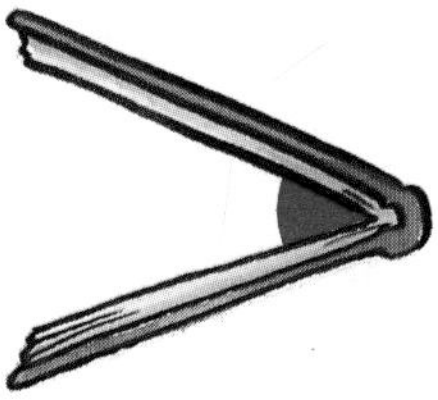

7

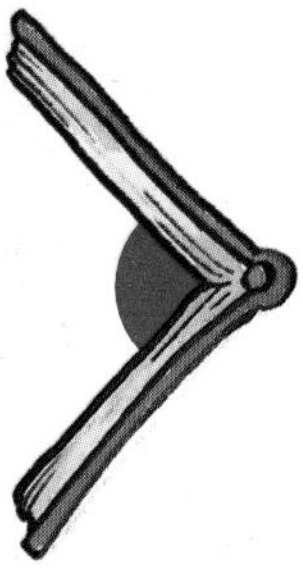

8

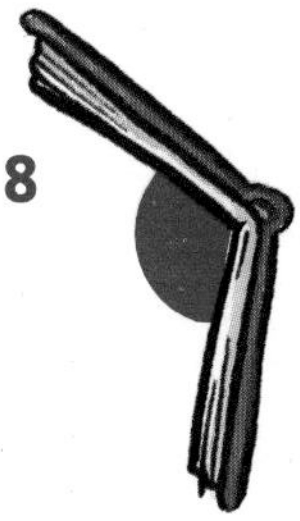

q

10

II

12

Copy these shapes and colour the right angles.

13

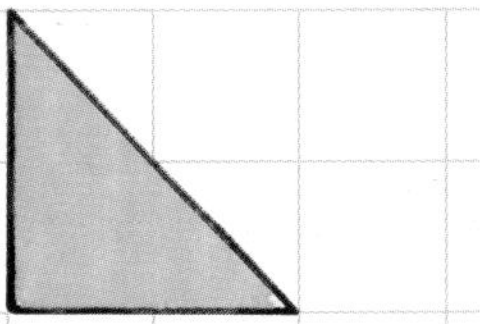

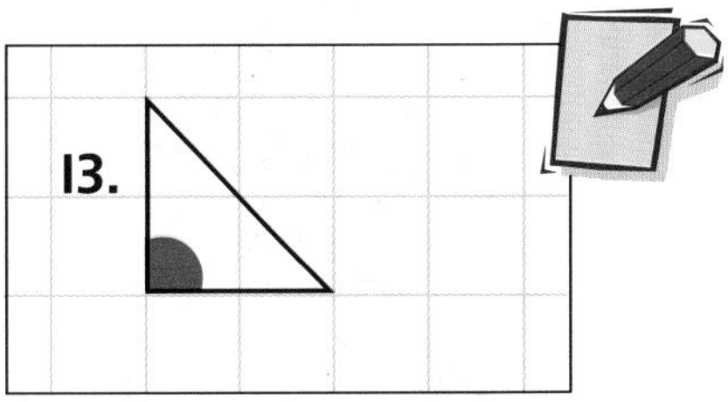

I3.

14

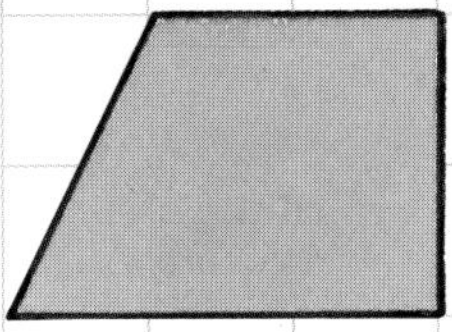

15

16

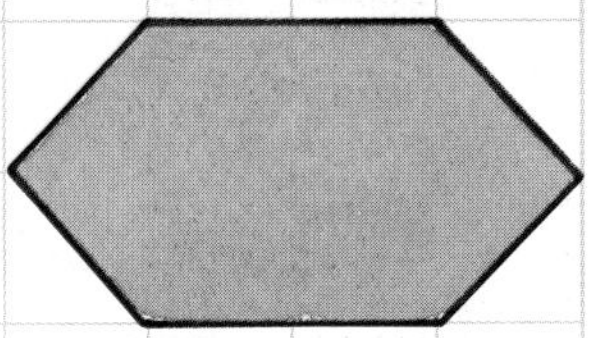

17

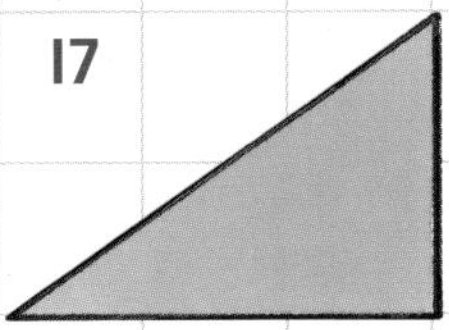

18

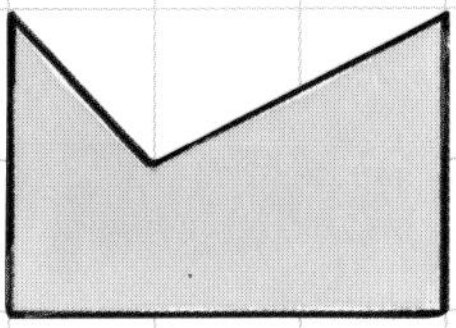

I9

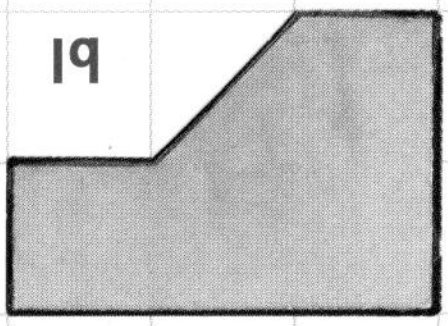

Right angle turns

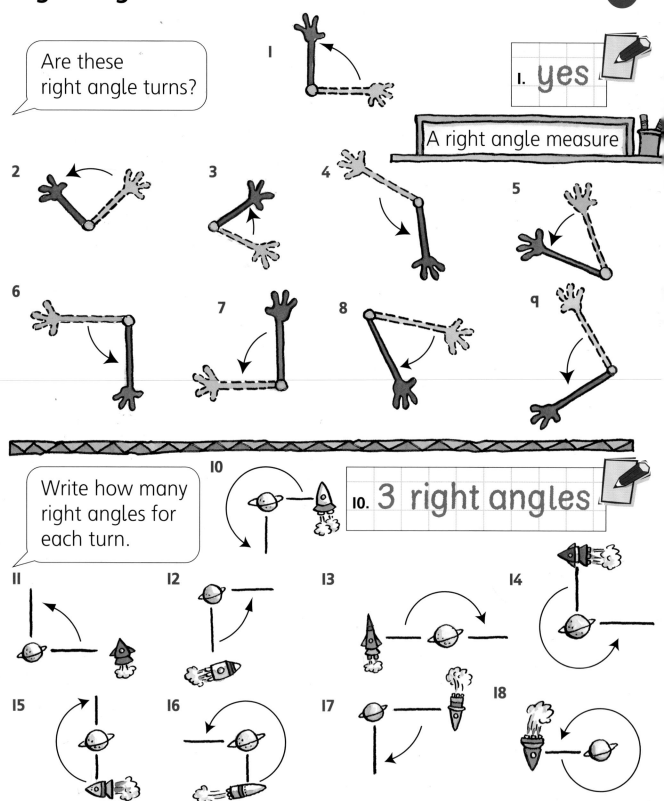

Are these right angle turns?

I. **yes**

A right angle measure

Write how many right angles for each turn.

10. **3 right angles**

e Which turns are clockwise, which are anticlockwise?

Right angle turns

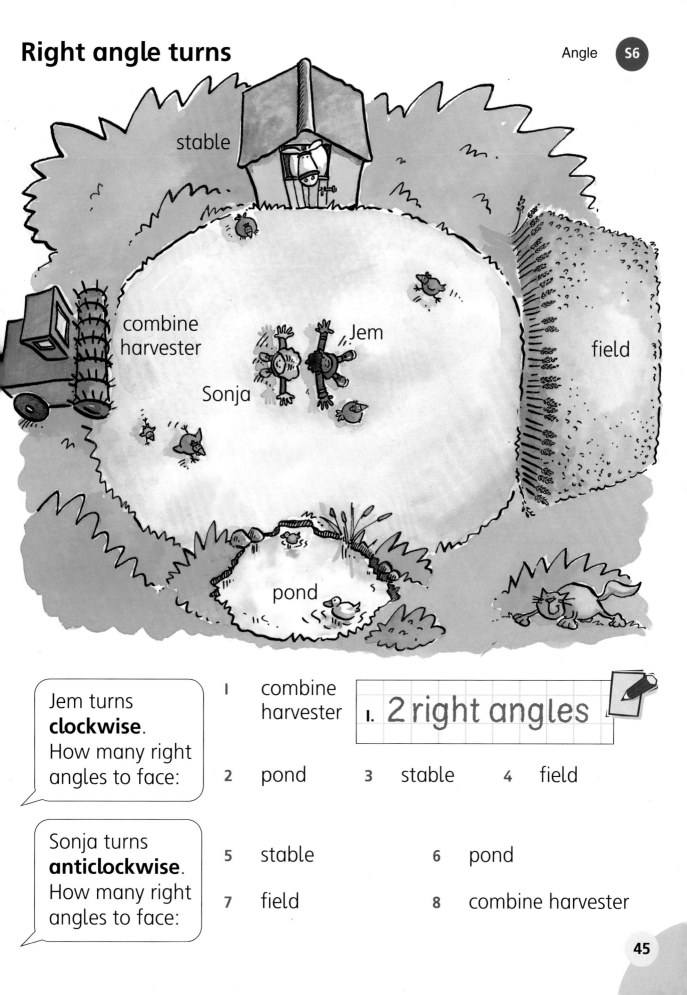

stable

combine harvester

Jem

Sonja

field

pond

Jem turns clockwise.
How many right angles to face:

1	combine harvester	1. 2 right angles

| 2 | pond | 3 | stable | 4 | field |

Sonja turns anticlockwise.
How many right angles to face:

| 5 | stable | 6 | pond |

| 7 | field | 8 | combine harvester |

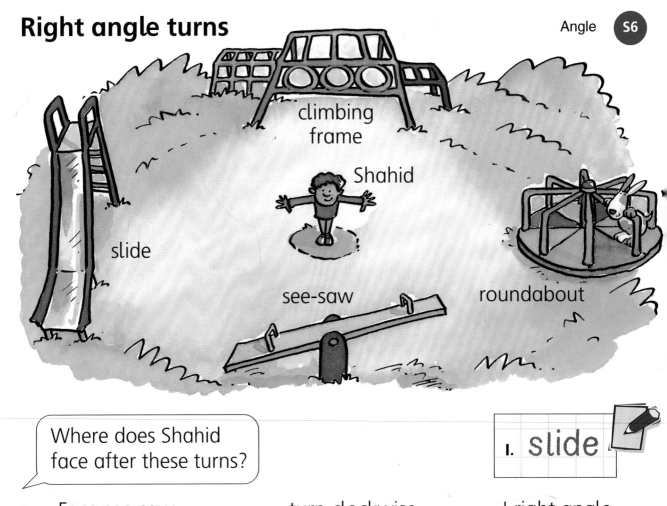

Where does Shahid face after these turns?

I. slide

I	Face see-saw	turn clockwise	I right angle
2	Face slide	turn anticlockwise	3 right angles
3	Face climbing frame	turn anticlockwise	I right angle
4	Face roundabout	turn clockwise	2 right angles
5	Face climbing frame	turn clockwise	I right angle
6	Face slide	turn clockwise	3 right angles
7	Face roundabout	turn anticlockwise	2 right angles
8	Face see-saw	turn anticlockwise	3 right angles

Write 5 different turns to finish facing the climbing frame.

Face roundabout, turn clockwise 3 right angles

Turning north, south, east or west

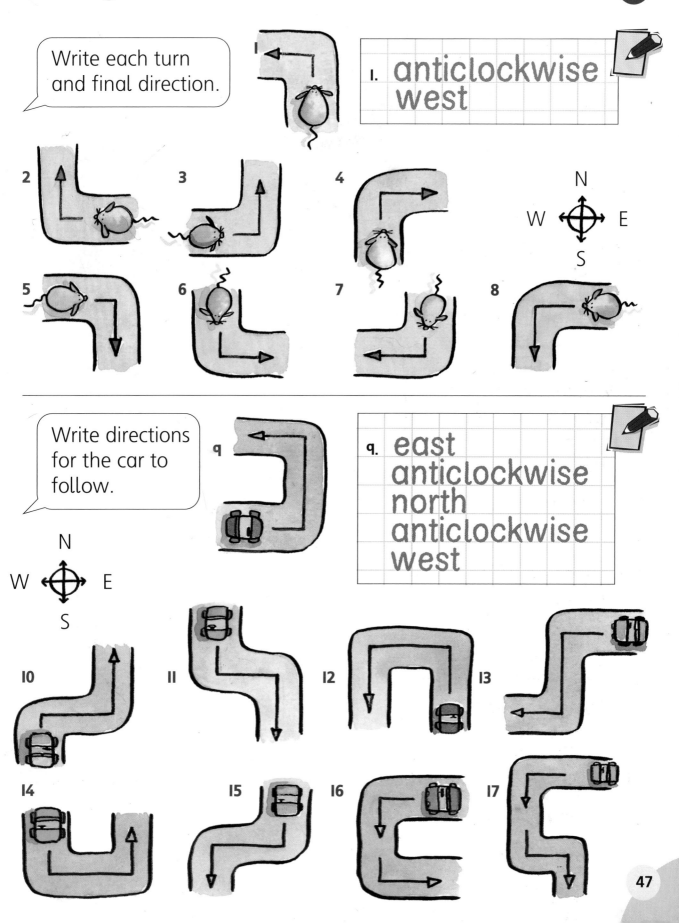

Write each turn and final direction.

1. anticlockwise
   west

N
W   E
S

Write directions for the car to follow.

N
W   E
S

9. east
   anticlockwise
   north
   anticlockwise
   west

Turning north, south, east or west

Write the direction before and after each turn.

I.

I. **east to south**

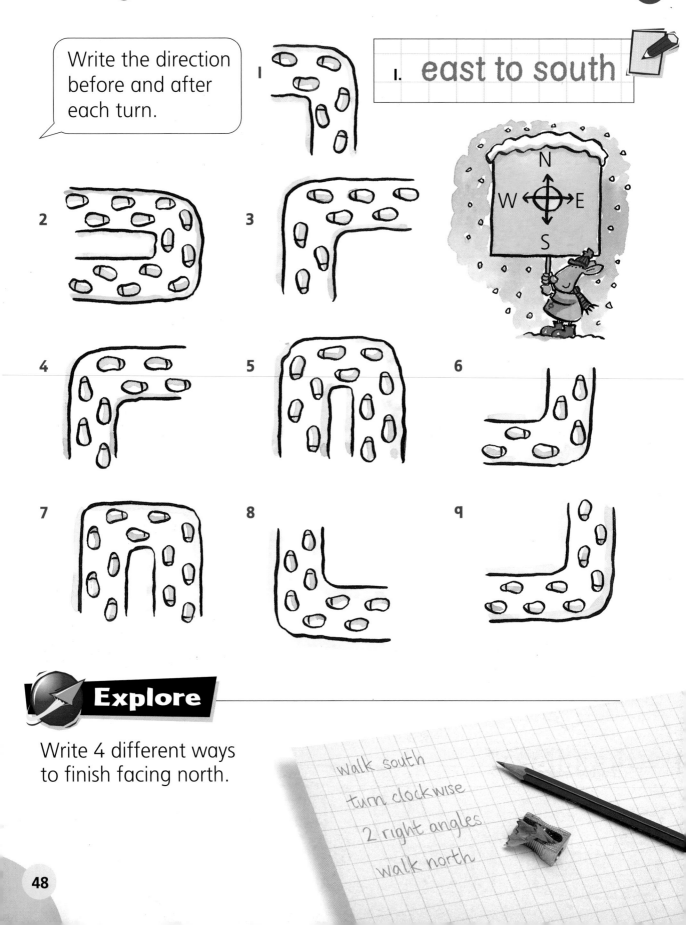

2

3

4

5

6

7

8

q

Explore

Write 4 different ways to finish facing north.

walk south

turn clockwise

2 right angles

walk north

Turning north, south, east or west

Write the direction before and after each turn.

1. I right angle

I. south to east

2. 3 right angles

3. 2 right angles

N
W ← → E
S

4. I right angle

5. 4 right angles

6. 2 right angles

7. 3 right angles

8. I right angle

9. 2 right angles

Write **clockwise** or **anticlockwise** for each turn above.

I. anticlockwise

Explore

Write 4 different turns that finish facing east.

face south
turn clockwise
I right angle
face east

Position

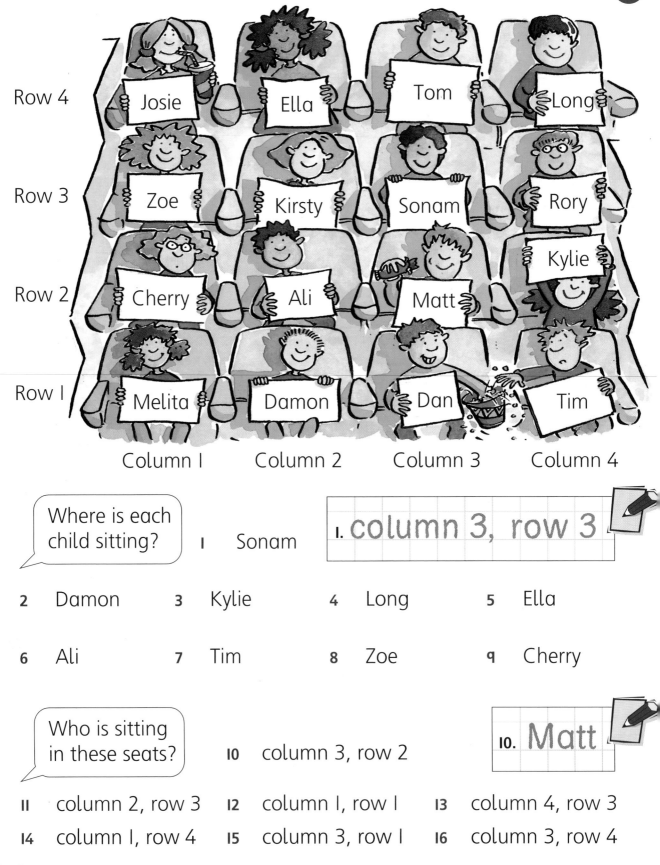

Row 4 — Josie, Ella, Tom, Long
Row 3 — Zoe, Kirsty, Sonam, Rory
Row 2 — Cherry, Ali, Matt, Kylie
Row 1 — Melita, Damon, Dan, Tim

Column 1   Column 2   Column 3   Column 4

Where is each child sitting?

1 Sonam

1. column 3, row 3

2 Damon   3 Kylie   4 Long   5 Ella

6 Ali   7 Tim   8 Zoe   9 Cherry

Who is sitting in these seats?

10 column 3, row 2

10. Matt

11 column 2, row 3   12 column 1, row 1   13 column 4, row 3

14 column 1, row 4   15 column 3, row 1   16 column 3, row 4

Position

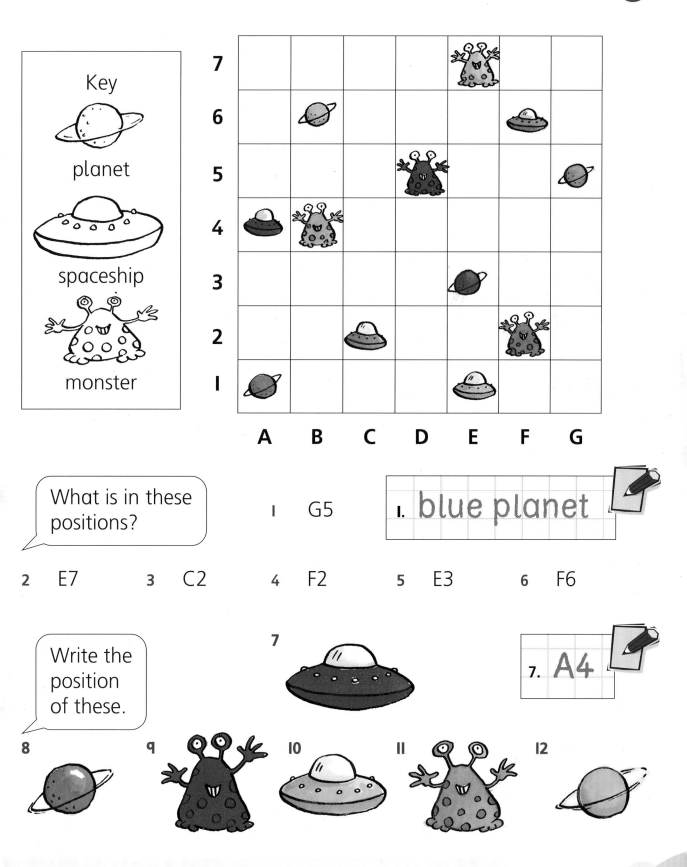

What is in these positions?

1 G5

I. blue planet

2 E7
3 C2
4 F2
5 E3
6 F6

Write the position of these.

7.

7. A4

8
9
10
11
12

ℯ Write the direction of different objects from the centre square (D4).

Position

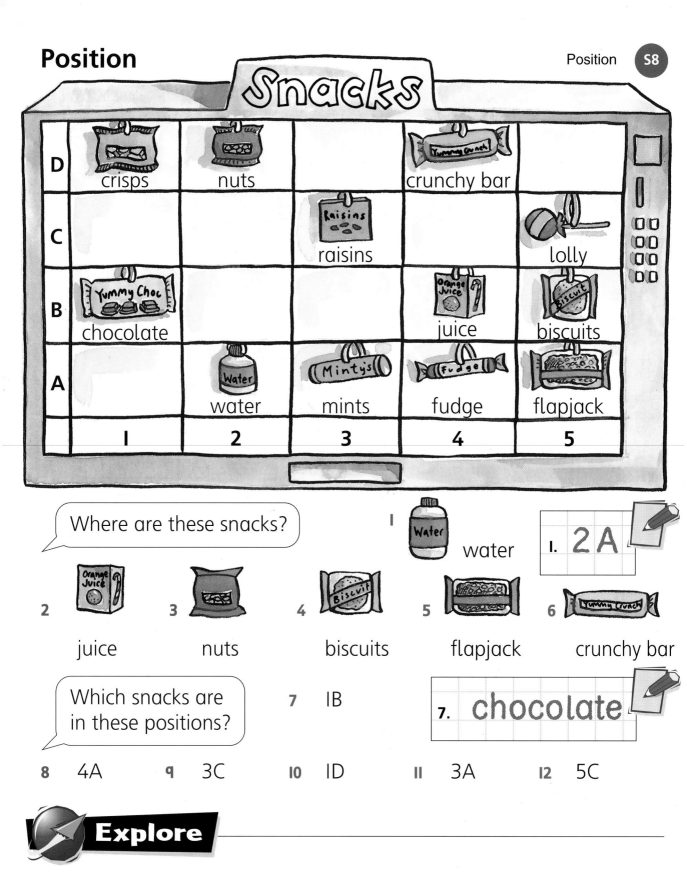

Snacks

	1	2	3	4	5
D	crisps	nuts		crunchy bar	
C			raisins		lolly
B	chocolate			juice	biscuits
A		water	mints	fudge	flapjack

Where are these snacks?

1 water

I. 2A

2 juice   3 nuts   4 biscuits   5 flapjack   6 crunchy bar

Which snacks are in these positions?

7 1B

7. chocolate

8 4A   9 3C   10 1D   11 3A   12 5C

Explore

Draw your own snack machine. Label the rows and columns.
Write the positions of the snacks.

52

Tally charts

> Copy the tally chart.
> Write all the totals.

Our favourite toys

Toys	Tallies	Total
yoyo	|||	3
ball	⅏ ⅏ ⅏ ||	
transformer	⅏ ⅏		
doll	⅏	
dinosaur	⅏ ||||	

> How many voted for:

1 doll? **2** ball? **3** dinosaur?

4 transformer? **5** yoyo?

> Which toy is:

6 most popular? **7** least popular?

> Which toys had:

8 more than 10 votes? **9** fewer than 8 votes?

> Here are 40 toys.

> Draw a tally chart to show how many there are of each type.

53

Tally charts

Draw a tally chart to show how many of each fruit there are.

Fruits in the picture

Fruits		Tallies
apples		
oranges		
bananas		
mangoes		
pineapples		
kiwis		

Copy this tally chart.

Draw tally marks to match the totals.

Our favourite colours

Colour	Tallies	Total
red		11
blue		21
yellow		7
purple		10
green		9
pink		4

Tally charts

Draw a tally chart to show how many caps there are of each colour.

@ Write the total for each colour.

Colours of the caps

Colour	Tallies
orange	
red	
yellow	
purple	
blue	
pink	

Explore

Throw a dice 30 times.

Draw a tally chart to show how many times you throw each number.

Repeat using two dice or a 10-sided dice.

Scores on the dice

Scores	Tallies	Total

Frequency tables

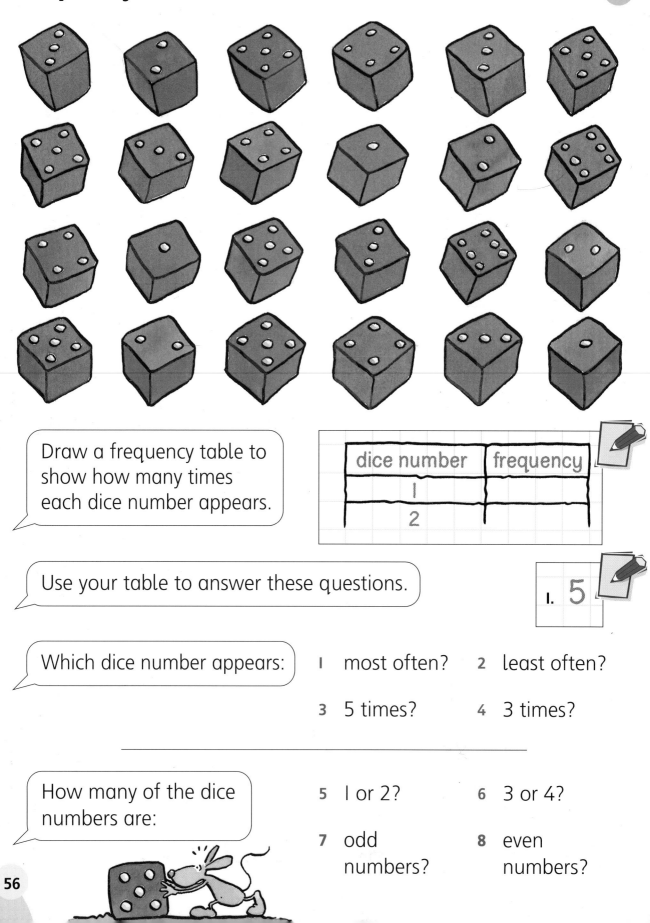

Draw a frequency table to show how many times each dice number appears.

dice number	frequency
1	
2	

Use your table to answer these questions.

1. 5

Which dice number appears:

1 most often?

2 least often?

3 5 times?

4 3 times?

How many of the dice numbers are:

5 1 or 2?

6 3 or 4?

7 odd numbers?

8 even numbers?

Frequency tables

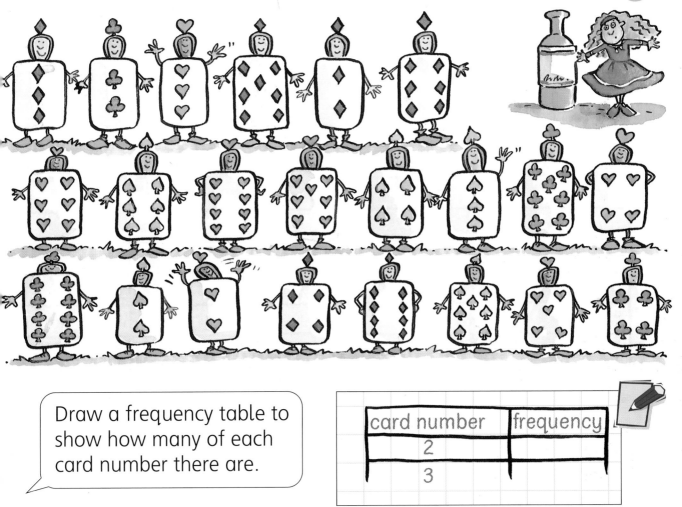

Draw a frequency table to show how many of each card number there are.

card number	frequency
2	
3	

Draw a frequency table to show how many of each suit there are.

suit		frequency
hearts	♥	
diamonds	♦	
clubs	♣	
spades	♠	

How many of these cards are:

1   4s?

2   6s?

3   3s?

4   5s?

5   3 or more?

6   less than 5?

7   odd?

8   even?

9   red?

10   black?

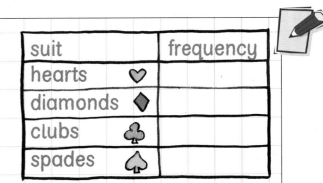

Frequency tables

The Year 3 children were asked their favourite sport. The votes were:

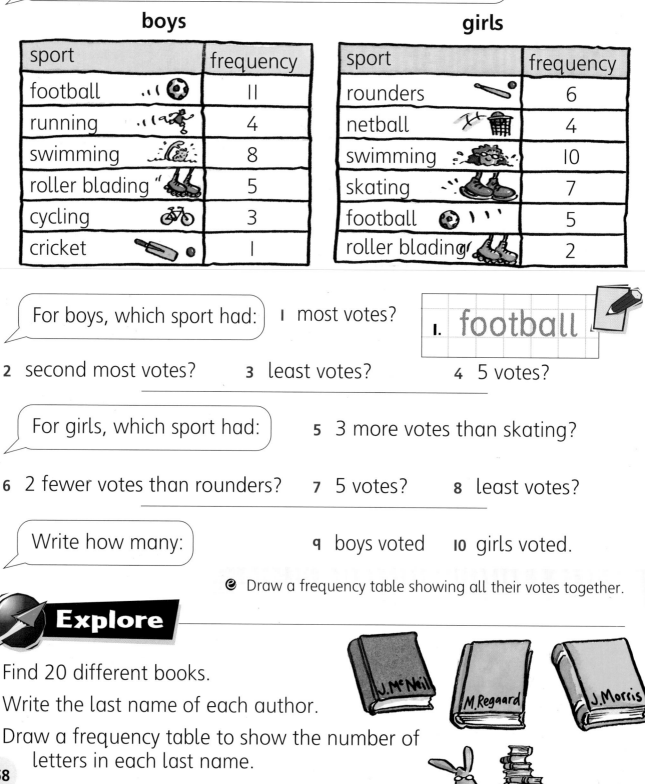

boys

sport		frequency
football		11
running		4
swimming		8
roller blading		5
cycling		3
cricket		1

girls

sport		frequency
rounders		6
netball		4
swimming		10
skating		7
football		5
roller blading		2

For boys, which sport had: **1** most votes?

1. football

2 second most votes? **3** least votes? **4** 5 votes?

For girls, which sport had: **5** 3 more votes than skating?

6 2 fewer votes than rounders? **7** 5 votes? **8** least votes?

Write how many: **9** boys voted **10** girls voted.

℮ Draw a frequency table showing all their votes together.

Explore

Find 20 different books.

Write the last name of each author.

Draw a frequency table to show the number of letters in each last name.

Bar graphs

Our favourite sports

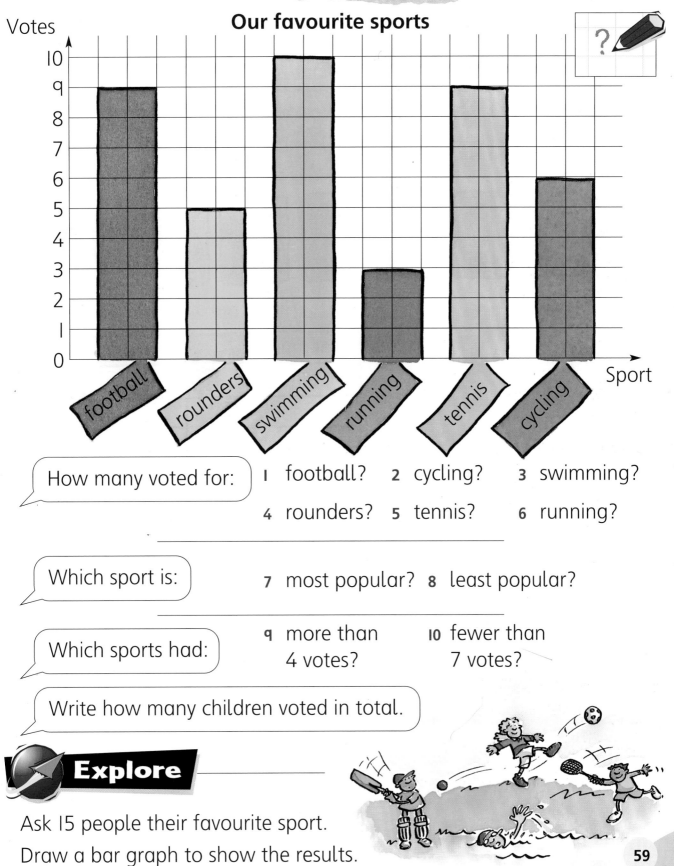

How many voted for:
1 football?   2 cycling?   3 swimming?
4 rounders?   5 tennis?   6 running?

Which sport is:
7 most popular?   8 least popular?

Which sports had:
9 more than 4 votes?   10 fewer than 7 votes?

Write how many children voted in total.

Explore

Ask 15 people their favourite sport.
Draw a bar graph to show the results.

Bar graphs

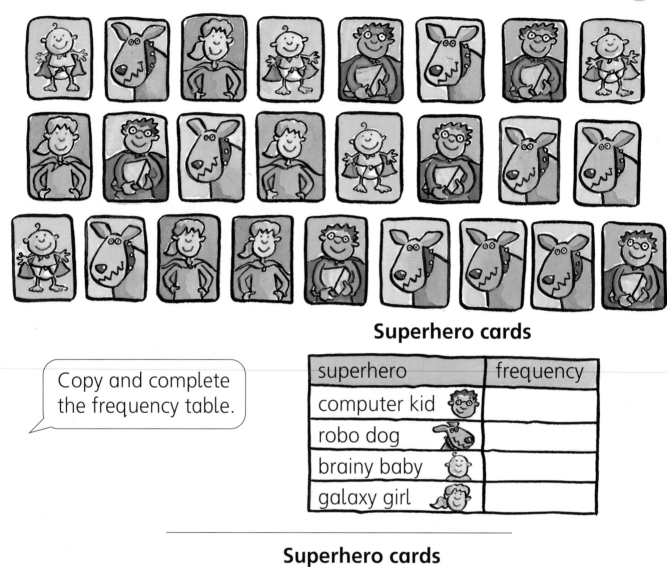

Superhero cards

Copy and complete the frequency table.

superhero		frequency
computer kid		
robo dog		
brainy baby		
galaxy girl		

Superhero cards

Copy and complete the bar graph.

Bar graphs

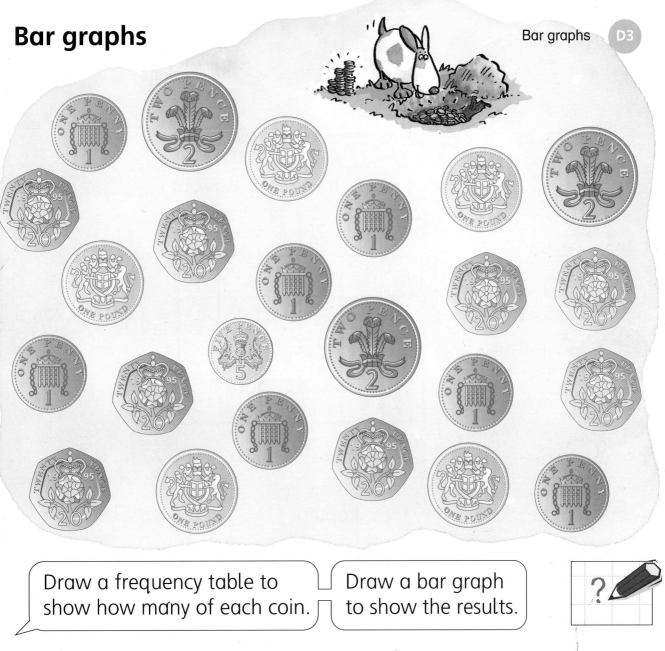

Draw a frequency table to show how many of each coin.

Draw a bar graph to show the results.

Explore

Work with a partner.

Talk about what you do in one day.

How many hours are you awake?

How many hours are you at school?

How many hours do you spend:

    eating? playing? watching TV?

Draw a bar graph to show the results.

Pictographs

Class F voted for their favourite ice-cream flavour. The votes were:

Favourite ice-cream flavours

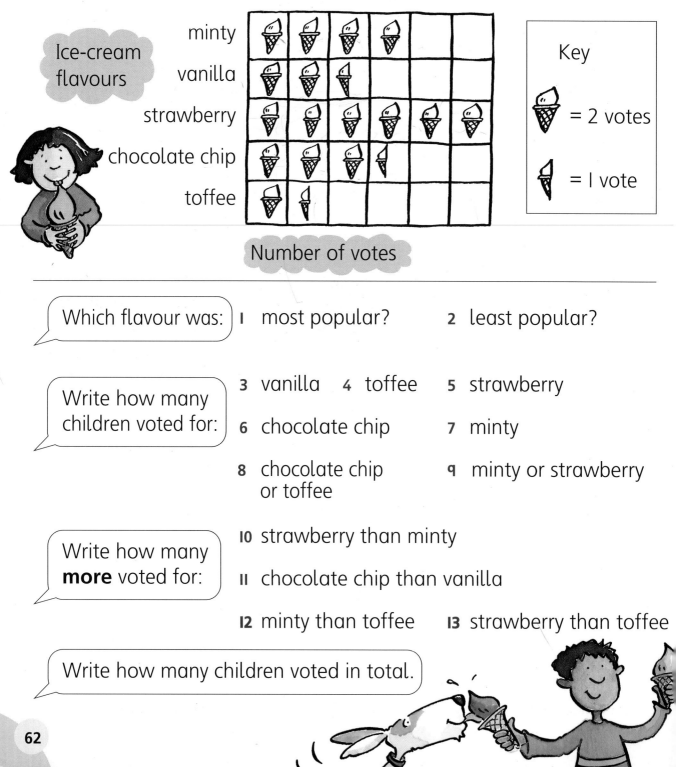

Ice-cream flavours

minty

vanilla

strawberry

chocolate chip

toffee

Number of votes

Key

= 2 votes

= 1 vote

Which flavour was: 1 most popular?   2 least popular?

Write how many children voted for:

3 vanilla   4 toffee   5 strawberry

6 chocolate chip   7 minty

8 chocolate chip or toffee   9 minty or strawberry

Write how many **more** voted for:

10 strawberry than minty

11 chocolate chip than vanilla

12 minty than toffee   13 strawberry than toffee

Write how many children voted in total.

Pictographs

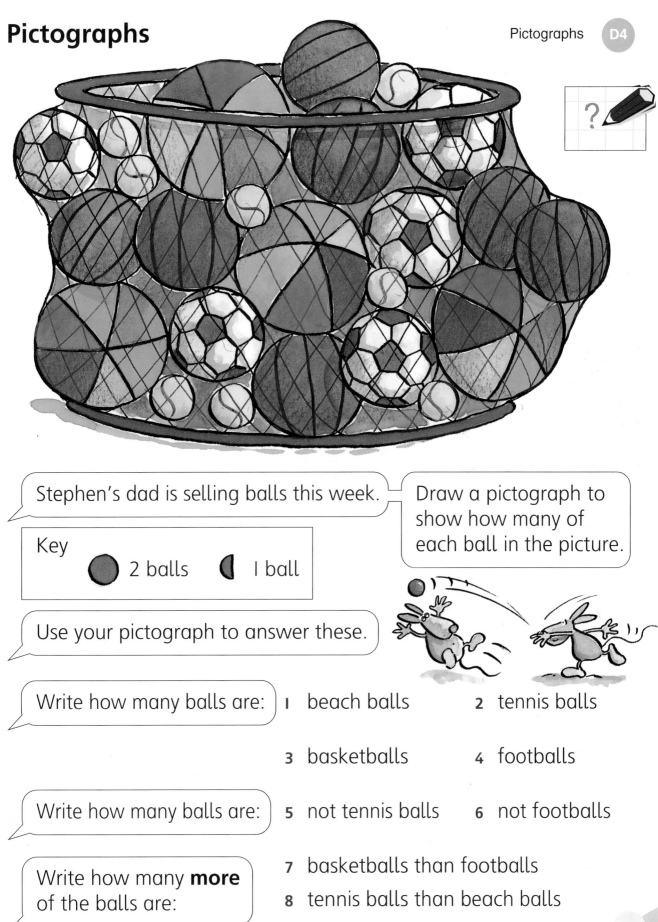

Stephen's dad is selling balls this week.

Draw a pictograph to show how many of each ball in the picture.

Key

⬤ 2 balls   ◖ I ball

Use your pictograph to answer these.

Write how many balls are:
1 beach balls
2 tennis balls

3 basketballs
4 footballs

Write how many balls are:
5 not tennis balls
6 not footballs

Write how many **more** of the balls are:
7 basketballs than footballs
8 tennis balls than beach balls

Pictographs

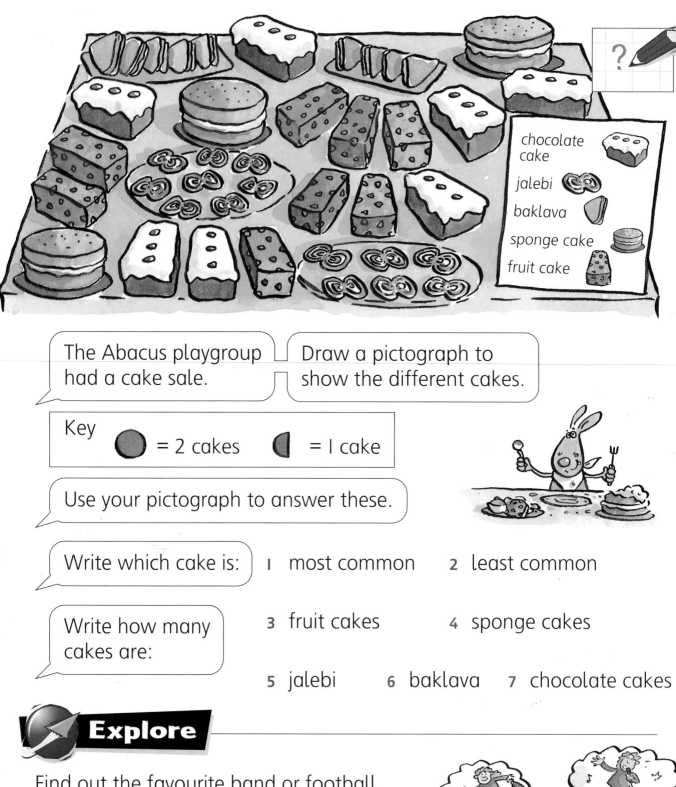

chocolate cake

jalebi

baklava

sponge cake

fruit cake

The Abacus playgroup had a cake sale.

Draw a pictograph to show the different cakes.

Key

⬤ = 2 cakes   ◖ = 1 cake

Use your pictograph to answer these.

Write which cake is: 1 most common   2 least common

Write how many cakes are:

3 fruit cakes   4 sponge cakes

5 jalebi   6 baklava   7 chocolate cakes

Explore

Find out the favourite band or football team of some children in school.

Draw a pictograph to show the results.